# MATH MADE EASY

A quick and easy guide to mental math and faster calculation.

by Reed Lerner

Copyright © 2016 Reed Lerner

All rights reserved

To my parents, for their unwavering support and unconditional love.

# INTRODUCTION

by Brett Moses

## "Pizza Problems"

A few weeks ago, I slipped away from the office for lunch. I dipped into the neighborhood pizzeria to find a young couple hunched over a greasy table, arguing. I'm not one to eavesdrop but, hey, I was curious. So I tuned in.

As it turned out, they were arguing about how much to tip the waiter. The man's iPhone had died, and his girlfriend had left hers at home, so they were left without any reasonable method of calculation.

Well, except for the human brain. And these days, who needs one of those?

As amusing as this scene was, it left me more than a little bit concerned. People have grown increasingly dependent on technology to solve even the simplest of our daily problems. Calculating a sale discount at the mall or splitting a check with friends should not require time wasted fumbling with a smartphone or finger counting.

I used to take my math skills for granted, but as I got older, I started to realize how much of a struggle math is for many people. After years of analyzing the way I process mental math, as well as learning various other techniques and tricks along the way, I've decided to create this short and easy guide to help others improve their math skills.

The idea is to teach you the easiest and most useful ways to do this; there are techniques that I have chosen not to include in order to prevent things from getting too dense and overly complicated.

# Who is this book for?

This book is for anyone looking to improve his or her ability to do math, from students to working professionals. Doing math keeps our minds stimulated (and even helps to prevent the onset of Alzheimer's), is practical and sometimes even necessary, and can save you time. Beyond that, it's part of being an intellectual person. You can learn to impress your teacher, your boss, your friends, or yourself with the techniques and tricks in this book. As of March 2016 the SAT will have math sections with no calculator allowed, making these skills more important than ever for students. There's something for those of you who say you're "terrible at math" as much as there is for the math whizzes. I've organized these techniques by complexity, and you should feel free to skip to those that you find most useful and interesting.

# TABLE OF CONTENTS

**PART I: THE BASICS ........................................................ 9**
   ADDITION & SUBTRACTION ................................... 10
      Addition ....................................................................... 10
      How to Do It Faster ................................................... 14
      Subtraction ................................................................. 17
      How to Do It Faster (Again) ..................................... 19
   MULTIPLICATION & DIVISION .................................. 22
      Multiplying & Dividing by Powers of 10 .................... 24
      Multiplying and Dividing by Multiples of 10 .............. 27

**PART II: A LITTLE SOMETHING FOR EVERYONE ..... 32**
   MORE MULTIPLICATION & DIVISION ...................... 33
      Numbers That End in 8 or 9 ....................................... 35
   Multiplying & Dividing by 5 .......................................... 38
      Multiplying & Dividing by Other Multiples of 5 .......... 41
      Multiplying by 11 ........................................................ 44
      The Double & Halve Technique for Multiplication .... 50
      Simplify to Make Division Easier ............................... 54
      Working with Percentages ......................................... 57

**Part III: FOR MATH LOVERS & STUDENTS ............... 64**
   TRICKS WITH 9 .......................................................... 65
      Dividing by 9 .............................................................. 65
      Turning Repeating Decimals into Fractions ............. 70

  SQUARES & SQUARE ROOTS ................................. 73
    Squaring Multiples of 5 .............................................. 73
  Squaring Other Numbers ............................................ 76
  Estimating Square Roots ............................................ 79

**APPENDIX** ........................................................................ **82**
  Terminology & Further Explanations ........................... 82
  Integers vs. Non-Integers ........................................... 82
  Order of Operations .................................................... 82
  Distributive Property of Multiplication ......................... 83
  Distributing Negatives ................................................. 84

**About The Author** ......................................................... **85**

# PART I: THE BASICS

Let's start nice and easy. In this section of the book we will review addition and subtraction techniques, as well as basic multiplication. If you're confident in these areas, you should go ahead and skip to PART II.

# ADDITION & SUBTRACTION

## Addition

Adding single digit numbers should be automatic. If you are a finger counter, take the time now to do some practice drills until you have the following table memorized.

| + | 0 | 1 | 2 | 3 | 4 | 5 | 6 | 7 | 8 | 9 |
|---|---|---|---|---|---|---|---|---|---|---|
| 0 | 0 | 1 | 2 | 3 | 4 | 5 | 6 | 7 | 8 | 9 |
| 1 | 1 | 2 | 3 | 4 | 5 | 6 | 7 | 8 | 9 | 10 |
| 2 | 2 | 3 | 4 | 5 | 6 | 7 | 8 | 9 | 10 | 11 |
| 3 | 3 | 4 | 5 | 6 | 7 | 8 | 9 | 10 | 11 | 12 |
| 4 | 4 | 5 | 6 | 7 | 8 | 9 | 10 | 11 | 12 | 13 |
| 5 | 5 | 6 | 7 | 8 | 9 | 10 | 11 | 12 | 13 | 14 |
| 6 | 6 | 7 | 8 | 9 | 10 | 11 | 12 | 13 | 14 | 15 |
| 7 | 7 | 8 | 9 | 10 | 11 | 12 | 13 | 14 | 15 | 16 |
| 8 | 8 | 9 | 10 | 11 | 12 | 13 | 14 | 15 | 16 | 17 |
| 9 | 9 | 10 | 11 | 12 | 13 | 14 | 15 | 16 | 17 | 18 |

## Adding Larger Numbers

Okay, now you can tell me that 7 + 8 = 15 without blinking. So why get tripped up adding 17 + 18? The key to adding larger numbers is to break them down into more manageable chunks.

| 1000s | 100s | 10s | 1s | . | 1/10s | 1/100s | 1/1000s |
|-------|------|-----|----|---|-------|--------|---------|
| 1 | 2 | 5 | 7 | . | 4 | 3 | 9 |

The **2** is in the hundreds place.
The **5** is in the tens place.
The **7** is in the ones, or units place.
The **4** is in the tenths place.
The **3** is in the hundredths place.
The **9** is in the thousandths place.

But why? Our number system is based on tens. It's probably no coincidence that we learn to count using our fingers and that we have ten of them. When we go past ten, we need to start again, so we say we have one ten and then add ones. What about when we get ten tens? Then we have one hundred. Ten hundreds become one thousand. And so on and so forth until we reach one googol and beyond. (No, I don't mean Google. One googol is a 1 followed by

100 zeros.) From our example above, let's drop the decimal and consider 1,257. It's made up of 7 ones, 5 tens, 2 hundreds, and 1 thousand.

$$1{,}257 = (1 \times 1{,}000) + (2 \times 100) + (5 \times 10) + (7 \times 1)$$

The 1 is called the thousands place because it represents the number of thousands, just as the 5 shows you how many tens there are.

As for the decimal part, .439 is 9 one thousandths ($1/1000$), 3 one hundredths ($1/100$), and 4 one tenths ($1/10$).

$$0.439 = (4 \times 1/10) + (3 \times 1/100) + (9 \times 1/1000)$$

This type of breakdown will be the key to how we attack many different kinds of problems.

Going back to 17 + 18 = ?

First, rewrite 17 as 10 + 7 and 18 as 10 + 8
The problem now becomes 17 + 18 = 10 + 7 + 10 + 8
Group the numbers based on place (ones, tens, hundreds, etc) to get
(10 + 10) + (7 + 8) = 20 + 15 = 35
Here's the same problem looking at just the math:

$$17 + 18 = ?$$
$$= 10 + 7 + 10 + 8$$
$$= (10 + 10) + (7 + 8)$$
$$= 20 + 15$$
$$= 35$$

For a second example, consider $65 + 73 = ?$ The goal is to see this problem as $(60 + 70) + (5 + 3) = ?$

$$65 + 73 = ?$$
$$= 60 + 5 + 70 + 3$$
$$= (60 + 70) + (5 + 3)$$
$$= 130 + 8$$
$$= 138$$

This technique works no matter how large the numbers are.

$$153 + 632 = ?$$
$$= (100 + 600) + (50 + 30) + (3 + 2)$$
$$= 700 + 80 + 5$$
$$= 785$$
$$435 + 568 = ?$$
$$= (400 + 500) + (30 + 60) + (5 + 8)$$
$$= 900 + 90 + 13$$
$$= 1{,}003$$

# How to Do It Faster

To solve problems even faster, and to make it easier to keep track of the numbers, apply this technique only to the number being added. Let's look at some of the same examples again.

$$17 + 18 = ?$$
$$= 17 + 10 + 8$$
$$= 27 + 8$$
$$= 35$$

$$65 + 73 = ?$$
$$= 65 + 70 + 3$$
$$= 135 + 3$$
$$= 138$$

$$153 + 632 = ?$$
$$= 153 + 600 + 30 + 2$$
$$= 753 + 30 + 2$$
$$= 783 + 2$$
$$= 785$$

Try some practice problems to make sure you've got it down.

## Add

1.  19     2.  64     3.  88     4.  72     5.  37
   + 24       + 28       + 45       + 49       + 75

6. 107    7. 243    8. 735    9. 518    10. 842
   + 48      + 155     + 265     + 328      + 280

## Answers

1. 43    2. 92    3. 133    4. 121    5. 112

6. 155   7. 398   8. 1,000   9. 846   10. 1,122

# Subtraction

*A note on negative numbers:*
*Adding a negative is the same as subtracting.*
*5 + -3 = 5 - 3 = 2*

We are going to approach subtraction in the same way we did addition.

For example, 23 - 18 = ?
First, rewrite 23 as 20 + 3 and 18 as 10 + 8
The problem now becomes
23 - 18 = (20 + 3) - (10 + 8)
Group based on place (ones, tens, hundreds, etc) to get (20 - 10) + (3 - 8) = 10 + -5 = 5

*Note that using the parentheses is necessary! If you want to understand more about order of operations and distributing negatives, check the appendix.*

Looking at just the math:

$$23 - 18 = ?$$
$$= (20 + 3) - (10 + 8)$$
$$= (20 - 10) + (3 - 8)$$
$$= 10 + -5$$
$$= 5$$

Let's try another:

$$54 - 12 = ?$$
$$= (50 - 10) + (4 - 2)$$
$$= 40 + 2$$
$$= 42$$

Let's try with bigger numbers. You may notice that I will skip the first step this time. With practice, you'll be able to look at your numbers and jump right to grouping them - I've used **bold** to help highlight this.

$$\mathbf{364} - 275 = ?$$
$$= (\mathbf{300} - 200) + (\mathbf{60} - 70) + (\mathbf{4} - 5)$$
$$= 100 + -10 + -1$$
$$= 89$$

$$\mathbf{8{,}372} - 5{,}498 = ?$$
$$= (\mathbf{8{,}000} - 5{,}000) + (\mathbf{300} - 400) + (\mathbf{70} - 90) + (\mathbf{2} - 8)$$
$$= 3{,}000 + -100 + -20 + -6$$
$$= 2{,}874$$

This last one is worth revisiting. With numbers that end in 8 or 9, we can round to the nearest multiple of 10. Let's break down 5,498 as 5,500 - 2 instead.

$$8{,}372 - 5{,}498 = ?$$
$$= (8{,}000 + 300 + 72) - (5{,}000 + 500 - 2)$$

$$= (8{,}000 - 5{,}000) + (300 - 500) + (72 - -2)$$
$$= (8{,}000 - 5{,}000) + (300 - 500) + (72 + 2)$$
$$= 3{,}000 + -200 + 74$$
$$= 2{,}874$$

*Notice that 72 - -2 became 72 + 2. Subtracting a negative is the same as adding a positive!*

If you are comfortable with this technique, it will often be faster. If you find it more confusing, just stick to the original.

## How to Do It Faster (Again)

Again, you can apply this technique only to the number being subtracted. Let's look at some of the same examples again.

$$23 - 18 = ?$$
$$= 23 - (10 + 8)$$
$$= 23 - 10 - 8$$
$$= 13 - 8$$
$$= 5$$

$$54 - 12 = ?$$
$$= 54 - (10 + 2)$$
$$= 54 - 10 - 2$$

= 44 - 2
= 42

Try skipping the first step.

364 - 275 = ?
= 364 - 200 - 70 - 5
= 164 - 70 - 5
= 94 - 5
= 89

Try some practice using whichever technique you prefer.

## Subtract

1. 78
   - 43
2. 64
   - 28
3. 89
   - 39
4. 94
   - 17
5. 75
   - 48
6. 107
   - 48
7. 243
   - 135
8. 495
   - 115
9. 519
   - 304
10. 738
    - 225

## Answers

1. 35  2. 36  3. 50  4. 77  5. 27
6. 59  7. 108  8. 380  9. 215  10. 513

# MULTIPLICATION & DIVISION

As you may have noticed, the key to making problems easier is breaking them down into more manageable chunks. We are going to apply the same idea going forward for multiplication and division with a heavy emphasis on 10 and multiples of 10. Let's get you comfortable working with these kinds of numbers first. Before getting started, you should have your basic multiplication table memorized.

| ×  | 0 | 1  | 2  | 3  | 4  | 5  | 6  | 7  | 8   | 9   | 10  | 11  | 12  |
|----|---|----|----|----|----|----|----|----|-----|-----|-----|-----|-----|
| 0  | 0 | 0  | 0  | 0  | 0  | 0  | 0  | 0  | 0   | 0   | 0   | 0   | 0   |
| 1  | 0 | 1  | 2  | 3  | 4  | 5  | 6  | 7  | 8   | 9   | 10  | 11  | 12  |
| 2  | 0 | 2  | 4  | 6  | 8  | 10 | 12 | 14 | 16  | 18  | 20  | 22  | 24  |
| 3  | 0 | 3  | 6  | 9  | 12 | 15 | 18 | 21 | 24  | 27  | 30  | 33  | 36  |
| 4  | 0 | 4  | 8  | 12 | 16 | 20 | 24 | 28 | 32  | 36  | 40  | 44  | 48  |
| 5  | 0 | 5  | 10 | 15 | 20 | 25 | 30 | 35 | 40  | 45  | 50  | 55  | 60  |
| 6  | 0 | 6  | 12 | 18 | 24 | 30 | 36 | 42 | 48  | 54  | 60  | 66  | 72  |
| 7  | 0 | 7  | 14 | 21 | 28 | 35 | 42 | 49 | 56  | 63  | 70  | 77  | 84  |
| 8  | 0 | 8  | 16 | 24 | 32 | 40 | 48 | 56 | 64  | 72  | 80  | 88  | 96  |
| 9  | 0 | 9  | 18 | 27 | 36 | 45 | 54 | 63 | 72  | 81  | 90  | 99  | 108 |
| 10 | 0 | 10 | 20 | 30 | 40 | 50 | 60 | 70 | 80  | 90  | 100 | 110 | 120 |
| 11 | 0 | 11 | 22 | 33 | 44 | 55 | 66 | 77 | 88  | 99  | 110 | 121 | 132 |
| 12 | 0 | 12 | 24 | 36 | 48 | 16 | 72 | 84 | 96  | 108 | 120 | 132 | 144 |

# Multiplying & Dividing by Powers of 10

Mathematically speaking, powers of 10 are numbers that can be represented as 10x where x is an integer.

In other words, powers of 10 are numbers like 10, 10,000, 0.1, 0.0001, etc. Essentially, any number that is a one followed by only zeros, or all zeros and then a one.

To multiply an integer by a power of 10, you can just add the number of zeros from your power of 10 onto the integer.

$$5 \times 1{,}000 = 5{,}000$$

$$18 \times 100 = 1{,}800$$

$$3 \times 10{,}000 = 30{,}000$$

Multiplying and dividing by powers of 10 is all about moving decimal places. For multiplication, we move the decimal to the right. Notice that when we were working with the integer examples above, the decimal was inferred.

From the example of 5 × 1,000 consider that 5 = 5.0

$5 \times 1{,}000 \rightarrow 5.0 \times 1{,}000 = 5{,}000.0$ (three zeros in 1,000 so the decimal moved three spots to the right)

Here are some examples with non-integer numbers:

$$5.5 \times 100 = 550.0$$

$$3.14159265 \times 1{,}000{,}000 = 3{,}141{,}592.65$$

Division works the same way, but we now move the decimal to the left.

$$5 \div 1000 = 0.005$$

$$5.5 \div 100 = 0.055$$

$$3{,}141 \div 100 = 31.41$$

Time to practice.

### Multiply & Divide by Powers of 10

1. 4 × 100
2. 8 × 1000
3. 0.45 × 10
4. 0.9 × 100
5. 3.4 × 100
6. 400 ÷ 100
7. 87 ÷ 1000
8. 6432 ÷ 10
9. 4.8 ÷ 100
10. 3 ÷ 1000

## Answers

1. 400   2. 8,000   3. 4.5   4. 90   5. 340
6. 4   7. 0.087   8. 643.2   9. 0.048   10. 0.003

# Multiplying and Dividing by Multiples of 10

When we work with multiples of 10 that are not necessarily a power (think 30, 110, anything that ends in one or more 0), we can still apply the same techniques.

First, ignore the zero(s) from the end of your number. Then do your multiplication or division ignoring the zeros.
Finally, move the decimal place based on the number of zeros.

$25 \times 30 = ?$
Ignore the 0 in 30 → $25 \times 3 = 75$
Then move the decimal one place to the right
→ 750

$128 \times 4,000 = ?$
Ignore the 0s in 4,000 → $128 \times 4 = (100 + 20 + 8) \times 4 = 400 + 80 + 32 = 512$
Then move the decimal three places to the right
→ 512,000

Let's look at some with just the math:

$$159 \times \mathbf{200} = ?$$
$$\to 159 \times 2 = ?$$
$$= 318 \to 31{,}8\mathbf{00}$$

$$9{,}435 \times 6{,}\mathbf{000} = ?$$
$$\to 9{,}435 \times 6 = ?$$
$$= (9{,}000 + 400 + 30 + 5) \times 6$$
$$= 54{,}000 + 2{,}400 + 180 + 30$$
$$= 56{,}610 \to 56{,}610{,}\mathbf{000}$$

You can even apply this technique when both numbers are multiples of 10.

7,400 × 2,000 = ?
Ignore the 0s in 7,400 AND 200 → 74 × 2 = 128
Then move the decimal **five** places to the right (two from 7,**400** and three from 2,**000**)
→ 12,800,000

Another example with just the math:

$$340 \times \mathbf{500} = ?$$
$$\to 34 \times 5 = ?$$
$$= 170 \to 170{,}\mathbf{000}$$

For division, the same rules apply but we must pay attention to which number is the multiple of 10. If we divide by a multiple of 10, we move the decimal left instead of right.

$48 \div 600 = ?$
Ignore the 0s in the 600 → $48 \div 6 = 8$
Then move the decimal two places to the left
→ 0.08

$135 \div 3,000 = ?$
Ignore the 0s in the 3,000 → $135 \div 3 = 45$
Then move the decimal three places to the left
→ 0.045

However if we divide a multiple of 10 by another number, we move the decimal right.

$4,800 \div 6 = ?$
Ignore the 0s in the 4,800 → $48 \div 6 = 8$
Then move the decimal two places to the right
→ 800

$135,000 \div 3$
Ignore the 0s in the 135,000: $135 \div 3 = 45$
Then move the decimal three places to the right
→ 45,000

What about when both numbers are multiples of 10? We will cover this later when we talk about simplification.

## Multiply & Divide by Multiples of 10

1. 3
  × 30

2. 7
  × 1500

3. 12
  × 800

4. 20
  × 450

5. 500
  × 400

6. 450
  ÷ 9

7. 2500
  ÷ 5

8. 56000
  ÷ 8

9. 48
  ÷ 600

10. 144
  ÷ 120

## Answers

1. 90  2. 10,500  3. 9,600  4. 9,000  5. 200,000

6. 50  7. 500  8. 7,000  9. 0.08  10. 1.2

# PART II: A LITTLE SOMETHING FOR EVERYONE

In this next section we will explore more multiplication and division. There are some general techniques as well as some specific tricks, and there's even a section devoted to percentages.

# MORE MULTIPLICATION & DIVISION

Now that you can comfortably work with multiples of 10, we will break down numbers the same way we did for addition and subtraction.

Let's start with examples where one of our numbers is a single digit.

5 × 16 = ?
Break the 16 into (10 + 6) → 5 × (10 + 6).
Next, distribute to get (5 × 10) + (5 × 6) = 50 + 30 = 80

*If you want to learn more about the distributive property of multiplication, see the appendix.*

$$7 \times 124 = ?$$
$$= 7 \times (100 + 20 + 4)$$
$$= 700 + 140 + 28$$
$$= 868$$

$$9 \times 128 = ?$$
$$= 9 \times (100 + 20 + 8)$$
$$= 900 + 180 + 72$$
$$= 1,152$$

When both numbers are more than one digit, we will split one number to turn the problem into two simpler multiplications, and then repeat the process from above if necessary.

$$24 \times 58 = ?$$
$$= 24 \times (50 + 8)$$
$$= (24 \times 50) + (24 \times 8)$$
$$= 1200 + (20 \times 8) + (4 \times 8)$$
$$= 1200 + 160 + 32$$
$$= 1392$$

$$16 \times 43 = ?$$
$$= (10 + 6) \times 43$$
$$= (10 \times 43) + (6 \times 43)$$
$$= 430 + (6 \times 40) + (6 \times 3)$$
$$= 430 + 240 + 18 = 688$$

## Numbers That End in 8 or 9

Sometimes it is advantageous to break down the numbers using subtraction instead of addition. This is especially the case with numbers ending in 9 because we will just multiply by a multiple of 10 and then subtract the original number. For example, if we are multiplying by 19 consider that 19 = 20 - 1

$$7 \times 19 = ?$$
$$= 7 \times (20 - 1)$$
$$= 140 - 7$$
$$= 133$$

$$13 \times 99 = ?$$
$$= 13 \times (100 - 1)$$
$$= 1300 - 13$$
$$= 1287$$

$$3 \times 78 = ?$$
$$= 3 \times (80 - 2)$$
$$= 240 - 6$$
$$= 234$$

$$12 \times 28 = ?$$
$$= 12 \times (30 - 2)$$
$$= 360 - 24$$
$$= 336$$

It's not a problem if you prefer to stick with the original technique using addition.

## Multiply by Other Numbers

1. 6
× 14

2. 7
× 36

3. 14
× 52

4. 6
× 132

5. 8
× 95

6. 24
× 99

7. 18
× 8

8. 12
× 29

9. 4
× 49

10. 9
× 124

## Answers

1. 84    2. 252   3. 728   4. 792   5. 760

6. 2,376  7. 144   8. 348   9. 196   10. 1,116

# Multiplying & Dividing by 5

If you can divide by powers of 10, you can divide by multiples of 5. It's all about relating the number you are dividing by to a power of 10. This will make more sense with some examples.

First, we want to relate 5 to a power of 10. $5 = 10 \div 2$, so when we multiply by 5 we can just multiply by 10 and then divide the answer by 2.

$$23 \times 5 = ?$$
$$= (23 \times 10) \div 2$$
$$= 230 \div 2$$
$$= 115$$

$$1{,}402 \times 5 = ?$$
$$= (1{,}402 \times 10) \div 2$$
$$= 14{,}020 \div 2$$
$$= 7{,}010$$

For division, remember that $5 = {}^{10}/_2$ so when we divide by 5, we are really multiplying by 2 and then dividing by 10. Confused? Let's refresh what happens when we divide by a fraction. The easiest way to do these problems is to take the reciprocal of the fraction (switch the top and the bottom) and then multiply instead of divide.

Say I want to do $8 \div 1/2 = ?$
First, recognize that the reciprocal of $1/2$ is $2/1$
This means $8 \div 1/2 = 8 \times 2/1 = 4$

Look at another example.

$$12 \div 3/4 = ?$$
$$\rightarrow 12 \times 4/3 = ?$$
$$= 12 \times 4 \div 3$$
$$= 48 \div 3$$
$$= 16$$

So, let's get back to dividing by 5.

$$65 \div 5 = ?$$
$$65 \div 10/2 = ?$$
$$\rightarrow 65 \times 2/10 = ?$$
$$= 65 \times 2 \div 10$$
$$= 130 \div 10$$
$$= 13$$

It's important to note that it's up to you whether you want to multiply by 2 and then divide by 10, or divide by 10 and then multiply by 2. In other words, $65 \times 2 \div 10 = 65 \div 10 \times 2$

$$140 \div 5 = ?$$
$$= (140 \div 10) \times 2$$
$$= 14 \times 2 = 28$$

$$85 \div 5 = ?$$
$$= (85 \times 2) \div 10$$
$$= 170 \div 10$$
$$= 17$$

# Multiplying & Dividing by Other Multiples of 5

You can now multiply and divide by any multiple of 5 simply by relating it to your answer when you multiply or divide by 5.

If you want to multiply by 15 for example, realize that 15 = 5 × 3. So simply multiply by 5, and then multiply by 3.

$$17 \times 15 = ?$$
$$= (17 \times 5) \times 3$$
$$= (17 \times 10) \div 2 \times 3$$
$$= 170 \div 2 \times 3$$
$$= 85 \times 3$$
$$= 255$$

120 ÷ 15 = ?
Consider that 15 is 5 × 3. Therefore we are going to divide by 5 and then by 3.
The answer will be (120 ÷ 5) ÷ 3 = 24 ÷ 3 = 8

140 ÷ 25 = ?
Consider that 25 is 5 × 5. Therefore we are going to divide by 5 and then again by 5.
The answer will be (140 ÷ 5) ÷ 5 = 28 ÷ 5 = 5 $^3/_5$

What if you don't want to divide by 5 first? You can always relate it directly to a division with a power of 10. To divide by 25, 25 = 100/4 so you can multiply by 4 then divide by 100. (Remember the whole reciprocal thing with switching top and bottom?)

$$140 \div 25 = ?$$
$$= (140 \times 4) \div 100$$
$$= 560 \div 100$$
$$= 5.6$$

To divide by 50, 50 is $^{100}/_2$ so multiply by 2 and then divide by 100.

$$750 \div 50 = ?$$
$$= (750 \times 2) \div 100$$
$$= 1500 \div 100$$
$$= 15$$

## Multiply and Divide by Multiples of 5

1. 5 × 18
2. 5 × 642
3. 15 × 8
4. 25 × 12
5. 25 × 60
6. 250 ÷ 5
7. 115 ÷ 5
8. 300 ÷ 5
9. 225 ÷ 25
10. 650 ÷ 50

## Answers

1. 90  2. 3,210  3. 270  4. 300  5. 1,500
6. 50  7. 23  8. 60  9. 9  10. 13

## Multiplying by 11

As you know from your times tables, multiplying 11 by a single digit number is just a matter of repeating the digit. 4 × 11 = 44, 8 × 11 = 88, etc.

When multiplying a two (or more) digit number by 11, it's not quite as easy, but there's a neat trick to help you out. Assume your answer will have the same first and last digits as your original number. To figure out what goes between them, we're going to add every pair of digits that are next to each other. It will make more sense after looking at some examples.
Let's follow the example of 26 × 11 = ? First, identify that the first digit of the answer will be 2 and the last will be 6 → 2 _ 6
Then, add them together to figure out the middle digit → 2 + 6 = 8
So 26 × 11 → 2 (2 + 6) 6 → 286

More examples will help make the process clear.

$$72 \times 11 = ?$$
$$72 \times 11 \rightarrow 7 \ (7 + 2) \ 2 \rightarrow 792$$
$$34 \times 11 = ?$$
$$34 \times 11 \rightarrow 3 \ (3 + 4) \ 4 \rightarrow 374$$

When the sum of the digits is 10 or more, we need to add the 1 to the digit on the left. Using the example of 48 × 11:

First, start by putting 4 as the first digit and 8 as the last → 4 _ 8
Next, add 4 + 8 to get 12. Instead of putting 12 between 4 and 8 to make 4,128 we add the 1 to the digit to its left (the 4) and keep only the 2 as the middle digit → (4 + 1) 2 8 → 528

48 × 11 → 4 (4 + 8) 8 → 4 12 8 → (4 + 1) 2 8 → 528

Again, try following along with a few more examples.

$$73 \times 11 = ?$$
$$73 \times 11 \rightarrow \underline{7} \; \underline{(7 + 3)} \; \underline{3} \rightarrow \underline{7} \; \underline{10} \; \underline{3} \rightarrow \underline{(7 + 1)} \; \underline{0} \; \underline{3} \rightarrow 803$$

$$59 \times 11 = ?$$
$$59 \times 11 \rightarrow 5 \; (5 + 9) \; 9 \rightarrow 5 \; 14 \; 9 \rightarrow (5 + 1) \; 4 \; 9 \rightarrow 649$$

Try this: 85 × 11 = ?

Did you get 935? If so, let's move on to three digit numbers. Remember, you will have to sum each pair of digits. In the example below, notice how the 4 is paired with the 1 to its left AND the 5 to its right.

$$145 \times 11 = ?$$
$$145 \times 11 \rightarrow 1 \ (1 + 4) \ (4 + 5) \ 5 \rightarrow 1{,}595$$

The same rule with sums of 10 or more still applies.

$$463 \times 11 = ?$$
$$463 \times 11 \rightarrow 4 \ (4 + 6) \ (6 + 3) \ 3 \rightarrow 4 \ 10 \ 9 \ 3$$
$$\rightarrow (4 + 1) \ 0 \ 9 \ 3 \rightarrow 5{,}093$$

What if more than one pair sums to greater than 10? Apply the same rule again, always adding the one to the digit immediately to the left in your answer.

$$578 \times 11 = ?$$
$$578 \times 11 \rightarrow 5 \ (5 + 7) \ (7 + 8) \ 8 \rightarrow 5 \ 12 \ 15 \ 8$$
$$\rightarrow 6 \ 2 \ 15 \ 8 \rightarrow 6{,}358$$

Here are some practice problems to make sure you've got it down.

## Multiply by 11

| 1. 11 | 2. 11 | 3. 11 | 4. 11 | 5. 11 |
|---|---|---|---|---|
| × 14 | × 35 | × 88 | × 54 | × 93 |

| 6. 11 | 7. 11 | 8. 11 | 9. 11 | 10. 11 |
|---|---|---|---|---|
| × 254 | × 145 | × 561 | × 183 | × 658 |

## Answers

1. 154   2. 385   3. 968   4. 594   5. 1,023
6. 2,794  7. 1,595  8. 6,171  9. 2,013  10. 7,238

Before we move on, let's look at some examples with even bigger numbers. Pay close attention to all of the digit pairings.

$$10{,}347 \times 11 = ?$$
$$10{,}347 \times 11 \rightarrow 1 \ (1+0) \ (0+3) \ (3+4) \ (4+7) \ 7$$
$$\rightarrow 1 \ 1 \ 3 \ 7 \ 11 \ 7 \rightarrow 113{,}817$$

The trickiest scenarios happen when you get a 9 followed by a pairing that adds to 10 or more. You will add the 1 to the 9, making it 10, and then will have to add the 1 from this ten to the digit to the left. For example:

$$4{,}638 \times 11 = ?$$
$$4{,}638 \times 11 \rightarrow 4 \ (4+6) \ (6+3) \ (3+8) \ 8 \rightarrow 4 \ 10 \ 9 \ 11 \ 8$$
$$\rightarrow 5 \ 0 \ 9 \ 11 \ 8 \rightarrow 5 \ 0 \ 10 \ 8 \rightarrow 51{,}018$$

$$5{,}729 \times 11 = ?$$
$$5{,}729 \times 11 \rightarrow 5 \ (5+7) \ (7+2) \ (2+9) \ 9 \rightarrow 5 \ 12 \ 9 \ 11 \ 9$$
$$\rightarrow 6 \ 2 \ 9 \ 11 \ 9 \rightarrow 6 \ 2 \ 10 \ 1 \ 9 \rightarrow 63{,}019$$

# The Double & Halve Technique for Multiplication

It's important to realize that we can change the way a problem looks without changing it's value. Multiplying by 1 is no different than multiplying by 9 and dividing by 9 because 9 ÷ 9 = 1. This is true for any real number because a number divided by itself is always 1.

Let's use this to our advantage to make problems easier. For multiplication problems, you will most often multiply one number by 2 and divide the other by 2. In other words, double one number and halve the other.

Most of us find multiplying by 2 to be far easier than multiplying by any greater number, so use double and halve to make one of your numbers a 2.

For example, to do 4 × 18 we will halve the 4 and double the 18 → 2 × 36 = 72

$$4 \times 18 = ?$$
$$= 4 \div 2 \times (18 \times 2)$$
$$= 2 \times 36$$
$$= 72$$

You can double and halve as many times as you would like.

$$43 \times 8 = ?$$
$$= (43 \times 2) \times 8 \div 2$$
$$= 86 \times 4$$
$$= (86 \times 2) \times 4 \div 2$$
$$= 172 \times 2$$
$$= 344$$

Double and halve can also be used to turn one of your numbers into a multiple of 10.

$$16 \times 15 = ?$$
$$= (16 \div 2) \times (15 \times 2)$$
$$= 8 \times 30$$
$$= 240$$

$$36 \times 25 = ?$$
$$= (36 \div 2) \times (25 \times 2)$$
$$= 18 \times 50$$
$$= (18 \div 2) \times (50 \times 2)$$
$$= 9 \times 100$$
$$= 900$$

It can be used to change your problem into any number you like working with. For example, to multiply by 11:

$$16 \times 22 = ?$$
$$= (16 \times 2) \times (22 \div 2)$$
$$= 32 \times 11$$
$$\rightarrow 3\ (3+2)\ 2 \rightarrow 352$$

## Multiply Using Double & Halve

| 1. 4 | 2. 18 | 3. 24 | 4. 5 | 5. 22 |
|---|---|---|---|---|
| × 24 | × 8 | × 15 | × 46 | × 26 |

## Answers

1. 96  2. 144  3. 360  4. 230  5. 572

## Simplify to Make Division Easier

Any division problem is essentially a fraction, so before attempting to divide you can always look to simplify. Simplification is when you divide both the numerator (top) and denominator (bottom) of a fraction by the same number. If you have $5/10$ you divide both by 5 to get $1/2$.

$460 \div 8 = ?$
Think of it as a fraction → $460/8 = ?$
Look for a common factor. Both are even so we can divide by 2 → $230/4$
Simplify more if possible, so divide top and bottom by 2 again → $115/2 = 57.5$

With multiples of 10, you can just cross out the same number of zeros from the top and bottom.

$54,000 \div 900 = ?$
→ $54,000/900 = ?$
→ $54,\cancel{000}/9\cancel{00} = 540/9 = 60$

## Simplify Before Dividing

1. 400 ÷ 20
2. 108 ÷ 4
3. 240 ÷ 8
4. 280 ÷ 35
5. 42 ÷ 6

## Answers

1. 20  2. 27  3. 30  4. 8  5. 7

## Working with Percentages

Percentages come up often - leaving tips at a restaurant, understanding statistics, and talking about business are just a few examples. Understanding them and how to do quick calculations is very important.
One thing not many people realize is that the percentage symbol "%" literally just means $1/100$ or $1 \div 100$.

Take an easy example. Everyone knows 50% of something is half, but why?

$50\% = 50(1/100) = 50/100 = 1/2$

The other important thing is to remember our lesson on dividing by powers of 10. In this case, we are dividing by 100 so we move the decimal two places to the left.

$1\% = 1 \div 100 = 0.01$
$10\% = 10 \div 100 = 0.10 = 0.1$
$50\% = 50.0 \div 100 = 0.50$
$6\% = 6.0 \div 100 = 0.06$
$135\% = 135.0 \div 100 = 1.35$

That means when you want 1% of something, just move the decimal two spots to the left. When you want 10%, just move it one spot.

$$10\% \text{ of } 350 = ?$$
$$\rightarrow 0.1 \times 350$$
$$= 35$$

So when you take 50% "of" something, you are really just multiplying by 50% (by half).

$$50\% \text{ of } 84 = ?$$
$$\rightarrow 0.5 \times 84$$
$$= 42$$

And when you take 25%, it's just a quarter.

$$25\% \text{ of } 360 = ?$$
$$\rightarrow {}^{25}/_{100} \times 360 = \tfrac{1}{4} \times 360$$
$$= 360 \div 4$$
$$= 90$$

Once this makes sense, we can use some of the same tricks from earlier in the book to make percentage questions easier. Let's say you want to leave a 15% tip on a bill that was $30.00
Using our technique from working with multiples of 5, we can break 15% into 10% + 5%.

10% is equal to .10 or $^1/_{10}$, so we just move the decimal one place to the left → $3
5% is half of 10%, so divide $3 by 2 → $1.50
Add them together to get $3 + $1.5 = $4.50

$$15\% \times 30 = ?$$
$$\to (10\% + 5\%) \times 30 = ?$$
$$= (10\% \times 30) + (5\% \times 30)$$
$$= (.10 \times 30) + (.05 \times 30)$$
$$= 3 + 1.5$$
$$= 4.5$$

How about a 20% tip? Well 20% is just twice 10%. So on the $30 bill, 20% would be 2 × (10% × 30) = 2 × 3 = $6.00

And 25%? 25% is just 5% more than 20%, so 2 × (10% × 30) + (5% × 30) = (2 × 3) + 1.5 = $7.50
You could also consider 25% as half of 50%. 50% × 30 = $^1/_2$ × 30 = 15 → $^1/_2$ × 15 = $7.50

Even when dealing with a more complicated number, the calculations are still pretty easy. Let's look at some examples for a bill of $84.20:

$$10\% \times 84.20 = ?$$
$$\rightarrow .10 \times 84.20$$
$$= 8.42$$

$$15\% \times 84.20 = ?$$
$$\rightarrow (10\% + 5\%) \times 84.20$$
$$= (10\% \times 84.20) + (5\% \times 84.20)$$
$$= (.10 \times 84.20) + (.05 \times 84.20)$$
$$= 8.42 + 4.21$$
$$= 12.63$$

$$20\% \times 84.20 = ?$$
$$(10\% \times 2) \times 84.20$$
$$= 2 \times (10\% \times 84.20)$$
$$= 2 \times (.10 \times 84.20)$$
$$= 2 \times 8.42$$
$$= 16.84$$

Here's another example: Harvard's most recent admission statistics say 6% of applicants were admitted. If 34,400 students applied, how many were admitted?

Consider that 6% is 5% + 1%. Let's find 5% by using the same technique of dividing 10% by 2. 1% is $1/100$ so we just move the decimal two spots to the left.

$$5\% \times 34{,}400 = ?$$
$$= (10\% \times 34{,}400) \div 2$$
$$= (0.1 \times 34{,}400) \div 2$$
$$= 3{,}440 \div 2$$
$$= 1{,}720$$

$$1\% \times 34{,}400 = ?$$
$$= 0.01 \times 34{,}400$$
$$= 344$$

$$6\% \times 34{,}400 = ?$$
$$= (5\% \times 34{,}400) + (1\% \times 34{,}400)$$
$$= 1{,}720 + 344$$
$$= 2{,}064$$

## Find the Percentages

1. 1% × 225
2. 10% × 93
3. 10% × 7
4. 15% × 85
5. 25% × 24
6. 50% × 82
7. 5% × 148
8. 5% × 420
9. 11% × 72
10. 6% × 28

Answers

1. 2.25  2. 9.3  3. 0.7  4. 12.75  5. 6
6. 41   7. 7.4  8. 21  9. 7.92   10. 1.68

# Part III: FOR MATH LOVERS & STUDENTS

We'll finish off with some neat tricks, however they're pretty specific and for most people won't come up too often in everyday life. For those of you that love math (or have started to while reading this book), are still in school, or are studying for admissions tests, let's learn a new way to divide by 9, the secret behind repeating decimals, and awesome techniques for squaring numbers and estimating square roots.

# TRICKS WITH 9

## Dividing by 9

Dividing by 9 uses a different technique with adding digits, so be careful not to confuse it with the one we learned for multiplying by 11. Let's walk through the steps while following an example of 12,321 ÷ 9 = ?
First, assume your answer will have the same first digit → 1
Next, add the next digit to your first digit → 1 + 2 = 3
→ This makes our current answer 1 3
To figure out what follows, add the next digit to the sum from the previous step (3).
→ 3 + 3 = 6 → new current answer 1 3 6
Repeat until the second to last digit of your original number → 6 + 2 = 8 → new current answer 1,368
Add last digit of original number to sum from previous step and divide by 9 → (8 + 1) ÷ 9 = 1
Add to your current answer → 1368 + 1 = 1,369

621 ÷ 9 = ?
Rewrite the first digit, then add with the next digit → 6 (6 + 2) → 6 8
Add last digit of original number to sum from previous step and divide by 9 → (8 + 1) ÷ 9 = 1
Add to your current answer → 68 + 1 = 69
More examples:

$$459 \div 9 = ?$$
$$\to 4 \ (4 + 5) \to 4 \ 9$$
$$\to 49 + (9 + 9) \div 9$$
$$= 49 + 2$$
$$= 51$$

$$4{,}230 \div 9 = ?$$
$$\to 4 \ (4 + 2) \to 4 \ 6 \ (6 + 3) \to 4 \ 6 \ 9$$
$$\to 469 + (9 + 0) \div 9$$
$$= 469 + 1$$
$$= 470$$

What about for a number that isn't a multiple of 9?
$145 \div 9 = ?$
Rewrite the first digit, then add with the next digit → 1 (1 + 4) → 15
Add last digit of original number to sum from previous step and divide by 9 → $(5 + 5) \div 9 = 10 \div 9 = 1\ ^{1}/_{9}$ or 1 remainder 1
Add to your current answer → $15 + 1\ ^{1}/_{9} = 16\ ^{1}/_{9}$

Another example:

$$2{,}321 \div 9 = ?$$
$$\to 2\ (2+3) \to 2\ 5\ (5+2) \to 2\ 5\ 7$$
$$\to 257 + (7+1) \div 9$$
$$= 257 + {}^8/_9$$
$$= 257\ {}^8/_9$$

If there are pairs that add up to 10 or more, we will carry (add) the tens place to the digit on the left.

$48{,}349 \div 9 = ?$
Rewrite the first digit, then add with the next digit $\to$ 4 (4 + 8) $\to$ 4  12 $\to$ 5  2
Add the next digit to the sum from the previous step $\to$ 5  2  (12 + 3) $\to$ 5  2  15 $\to$ 5  3  5
Repeat until the second to last digit of your original number $\to$ 5  3  5  (15 + 4) $\to$ 5  3  5  19 $\to$ 5,369
Add last digit of original number to sum from previous step and divide by 9 $\to$ (19 + 9) $\div$ 9 = 28 $\div$ 9 = $3{}^1/_9$
Add to your current answer $\to$ 5,369 + $3{}^1/_9$ = $5{,}372{}^1/_9$

Another example:

$$1{,}475 \div 9 = ?$$
$$\to 1\ (1+4) \to 1\ 5\ (5+7)$$
$$\to 1\ 5\ 12 \to 1\ 6\ 2$$
$$\to 162 + (12 + 5) \div 9$$
$$= 162 + 1\,{}^8/_9$$
$$= 163\,{}^8/_9$$

$$54{,}832 \div 9 = ?$$
$$\to 5\ (5+4) \to 5\ 9\ (9+8) \to 5\ 9\ 17\ (17+3) \to 5\ 10\ 7\ (17+3)$$
$$\to 6\ 0\ 7\ 20 \to 6\ 0\ 9\ 0$$
$$\to 6{,}090 + (20 + 2) \div 9$$
$$= 6{,}090 + {}^{22}/_9$$
$$= 6{,}090 + 2\,{}^4/_9$$
$$= 6{,}092\,{}^4/_9$$

Divide by 9

1. 279 ÷ 9
2. 684 ÷ 9
3. 351 ÷ 9
4. 918 ÷ 9
5. 414 ÷ 9
6. 523 ÷ 9
7. 839 ÷ 9
8. 1,023 ÷ 9
9. 2,371 ÷ 9
10. 1,873 ÷ 9

## Answers

1. 31     2. 76     3. 39     4. 102     5. 46

6. $58^1/_9$   7. $93^2/_9$   8. $113^6/_9$   9. $263^4/_9$   10. $208^1/_9$

## Turning Repeating Decimals into Fractions

Any repeating decimal can easily be turned into a fraction. First, identify the number that is repeating and note how many digits there are in this repetition. The numerator of the fraction will be the repeating number from the decimal. The denominator will be composed of 9s, (numbers like 9, 99, 999, 9999, etc.) depending on how many digits there are in the original repetition.

For example, 0.25252525…. is just 25 repeating (two digits), therefore the answer will be $^{25}/_{99}$.

More examples:

0.1111… = $^{1}/_{9}$

0.645645645 = $^{645}/_{999}$

It holds true for some of the most common repeating decimals, like 0.333…
0.3333 = $^{3}/_{9}$ = $^{1}/_{3}$

## Convert to Fraction Form

1. 0.666…
2. 0.2424…
3. 0.532…
4. 0.1111…
5. 0.6767…

## Answers

1. $6/9 = 2/3$  2. $24/99$  3. $532/999$  4. $1/9$  5. $67/99$

# SQUARES & SQUARE ROOTS

## Squaring Multiples of 5

To find the square of any multiple of 5 is easy. First, cut the 5 off your number and multiply the remaining number by one more than itself. Then just put a 25 on to the end - your answer will always end in 25.

$$35^2 = ?$$
$$3\cancel{5} \to 3 \times (3 + 1)$$
$$= 3 \times 4$$
$$= 12$$
$$\to 1,225$$

Try skipping the first step.

$$75^2 = ?$$
$$7\cancel{5} \to 7 \times 8$$
$$= 56$$
$$\to 5,625$$

Let's try a three digit number.

$$105^2 = ?$$
$$10\cancel{5} \to 10 \times 11$$
$$= 110$$
$$\to 11,025$$

Try some more now.

## Square

1. $45^2$  2. $65^2$  3. $95^2$  4. $115^2$  5. $205^2$

## Answers

1. 2,025   2. 4,225   3. 9,025   4. 13,225   5. 42,025

## Squaring Other Numbers

You won't always be looking at numbers that are multiples of 5, so let's look at a technique that works for all numbers. Using $19^2$ as an example:
1) First, find the difference to the nearest multiple of 10. Let's call this $d$. → $d = 20 - 19 = 1$
2) Next, multiply (number + $d$) × (number - $d$).
→ $(19 + 1) \times (19 - 1) = 20 \times 18 = 360$
3) Square $d$ and add it to your answer from step 2 (360). → $1^2 = 1$ → $360 + 1 = 361$

Follow along with some examples.

$16^2 = ?$
1) Nearest multiple of 10 is 20 → $d = 20 - 16 = 4$
2) $(16 + 4) \times (16 - 4) = 20 \times 12 = 240$
3) $d^2 = 4^2 = 16$
→ $240 + 16 = 256$

$28^2 = ?$
1) Nearest multiple of 10 is 30 → $d = 30 - 28 = 2$
2) $(28 + 2) \times (28 - 2) = 30 \times 26 = 780$
3) $d^2 = 2^2 = 4$
→ $780 + 4 = 784$

## Square

1. $29^2$  2. $32^2$  3. $99^2$  4. $17^2$  5. $46^2$

## Answers

1. 841  2. 1,024  3. 9,801  4. 289  5. 2,116

# Estimating Square Roots

This technique is used to estimate the square root of non-perfect squares. It's used to work within the range of the squares you know, so we'll assume you know up to $15^2 = 225$. Let's walk through the process using 86 as an example:

$\sqrt{86} = ?$
1) Get the closest square → 81
2) Divide 86 by the square root of the closest square → $86 \div 9 = 9\,{}^5/_9 = 9.555\ldots$
3) Average this number and the square root of the closest square: $(9 + 9.555) \div 2 \rightarrow 9\,{}^{2.5}/_9 = 9.252525\ldots$
4) 9.273 is the real answer. The closer your chosen square is, the more accurate your estimate will be.

A few more examples.

$\sqrt{70} = ?$
Closest square: 64
Divide original number by the square root of closest square → $70 \div 8 = 8\,{}^6/_8$
Average with square root of closest square → $8\,{}^3/_8 = 8.375$
Compare to real square root: 8.367

$\sqrt{7}$ = ?

Closest square: 9

Divide original number by the square root of closest square → 7 ÷ 3 = 2 $\frac{1}{3}$

Average with square root of closest square → 2 $\frac{2}{3}$ = 2.667

Compare to real square root of 2.65

## Estimate The Square Root

1. $\sqrt{34}$  2. $\sqrt{80}$  3. $\sqrt{18}$  4. $\sqrt{115}$  5. $\sqrt{200}$

## Answers

1. $5^5/_6$  2. $8^{8.5}/_9$  3. $4^1/_4$  4. $10^8/_{11}$  5. $14^2/_{14}$

# APPENDIX

## Terminology & Further Explanations

If some of the math terminology left you scratching your head, here's the place to clear things up.

## Integers vs. Non-Integers

An integer is a number without a fractional or decimal component. 5, 984, 0, and -1,829 are all integers. 1.4, $5^1/_2$, -7.2, and are not. Why didn't I just say "an integer is a whole number"? Whole numbers, also known as natural numbers, don't include the negatives (or zero, depending on who you ask).

## Order of Operations

The order of operations tells you how to perform a calculation. **P**arentheses → **E**xponents → **M**ultiplication/**D**ivision → **A**ddition/**S**ubtraction, or **PEMDAS** for short.

# Distributive Property of Multiplication

The distributive property states that
$a \times (b + c) = (a \times b) + (a \times c)$

Let's look at an example similar to those found in the book, $8 \times (10 + 3) = ?$

We have to perform the operation in the parentheses first, so it becomes $8 \times 13 = 104$
If we had ignored the order of operations and had done just left to right, it would have become
$8 \times 10 + 3 = 80 + 3 = 83$

This may feel a little backwards, since earlier we were taking $8 \times 13 = ?$ and breaking it into
$8 \times (10 + 3) = ?$

## Distributing Negatives

Distributing a negative is a specific example of the distributive property of multiplication.

$(a + b) - (c + d) = a + b - c - d$

Consider that subtracting is the same as adding a negative.

$(a + b) - (c + d) = a + b + (-1)(c + d)$

$a + b + (-1)(c + d) = a + b + (-1)(c) + (-1)(d)$
$= a + b - c - d$

Here it is applied to an example similar to those in the subtraction explanation from Part I.

$$28 - 13 = ?$$
$$= (20 + 8) - (10 + 3)$$
$$= (20 + 8) + (-1)(10 + 3)$$
$$= (20 + 8) + (-1)(10) + (-1)(3)$$
$$= 20 + 8 - 10 - 3$$
$$= 15$$

# About The Author

Reed Lerner started tutoring in high school and has since taught students from around the world in math, physics, and an array of standardized tests including the SAT, GMAT and ISEE. Born in Los Angeles, California, he attended the University of Pennsylvania and graduated with degrees in mechanical engineering and marketing.

Reed is available for private tutoring in person and online. You can reach him via email (reed.lerner@gmail.com) to inquire further.

Made in the USA
Lexington, KY
12 February 2017